Words in **bold** are
explained in the
glossary on page 31.

What are flies and mosquitoes?

Flies and mosquitoes are **insects**. Their bodies are divided into three parts. They have a head with a pair of eyes and **antennae**, a thorax or chest with six legs and a back part called an **abdomen**. These insects have one pair of wings which are attached to the thorax.

Mosquitoes have six long legs attached to their thorax.

Looking at Minibeasts

Flies and Mosquitoes

Sally Morgan

Contents

A mosquito is a biting insect. Its mouthparts are sharp enough to pierce skin.

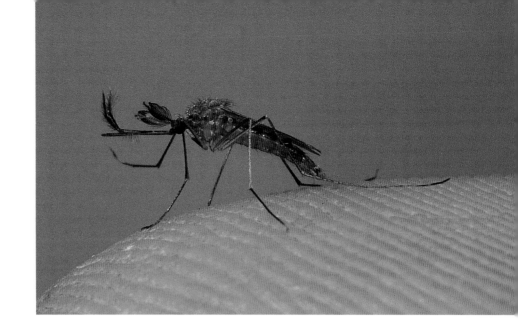

Large flies, such as this housefly, have hairy bodies. They have a large pair of eyes and a short pair of antennae.

The fly family

There are more than 90 000 different types of fly. They range in size from large horseflies to tiny midges. They are found all over the world, from very hot to very cold countries.

This is a giant fly. It is about 2 centimetres long and is the largest type of fly in the world.

Many flies are found around flowers, feeding on nectar.

Most insects have two pairs of wings, but mosquitoes and flies only have one pair. The second pair has been replaced by two knobbed stalks called **halteres**. An insect such as a butterfly has 'fly' in its name, but is not a true fly because it has four wings.

This long-legged fly is called a mimegralla. It is a true fly because it only has two wings.

Seeing and feeling

A fly has two large compound eyes. If you look very carefully at the eye of a fly you may be able to see that it is made up of many mini-eyes.

This close up shows how each eye is made up of mini-eyes arranged in rows.

A compound eye sees a picture that is made up of lots of tiny images. This means that flies cannot see things in detail, but they can see movement.

The eyes of this fly are on the end of long stalks. They look like a pair of antennae.

Flies use their antennae, or feelers, to find food. Their antennae are short and hairy. They taste their food using their feet. Their feet are covered in taste buds – just like our tongues.

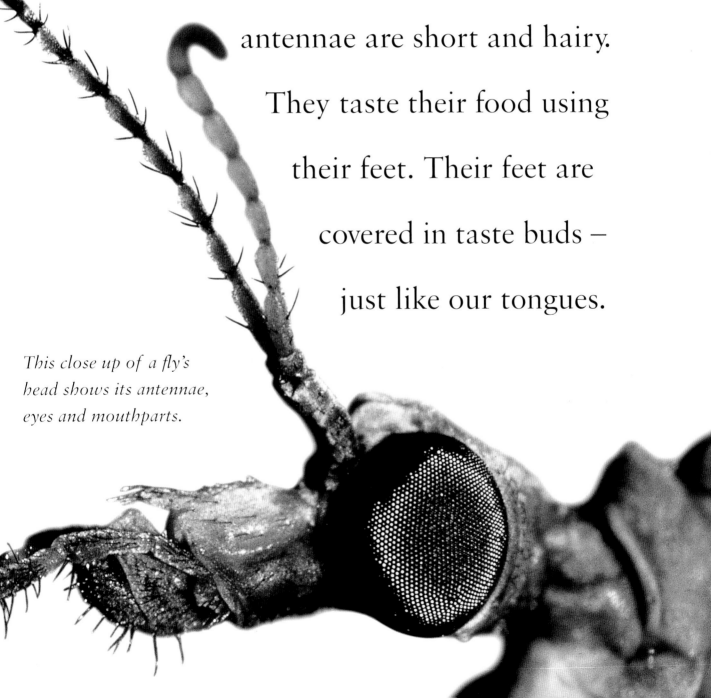

This close up of a fly's head shows its antennae, eyes and mouthparts.

Thin wings

The wings of a fly are thin and usually transparent. This means you can see through them. Flies can move their wings up and down very quickly. Some of the smallest flies beat their wings thousands of times each second.

A crane fly has long, delicate wings which are easily damaged.

This fly's long legs hang down as it flies around.

Hover flies can hover in front of a flower without moving.

The wings beat so fast that they make a buzz or whine. People know that mosquitoes are about at night because they can hear a whining sound as they fly around the room.

Sucking and biting

A fly has a mouth that can suck up food. It is called a **proboscis** and it looks like a tube with two suckers at the end. When a fly lands on food, it pours out the contents of its stomach. This turns the food into a liquid which the fly can suck up. Mosquitoes feed on blood. They have long, sharp mouthparts, which they use to pierce skin.

A mosquito pierces the skin with its mouthparts, then sucks up the blood.

Hover flies suck up the sugary
nectar found in flowers.

This empid fly is feeding
on a moth it has caught.

Houseflies

Many flies are found in and around the home. They are attracted to the food in the kitchen and rubbish bin. There are more flies around during the warm, summer months. Houseflies often have brightly-coloured bodies. Bluebottles and greenbottles are types of housefly.

On sunny days, the greenbottle can be seen buzzing around rubbish bins.

Bluebottles or blowflies are large, hairy, metallic-blue flies that buzz loudly. Here they are feeding on some rotting food.

Some flies spend the winter in our homes. They creep into warm, dry places in attics and walls.

Yellow swarming flies may gather in large groups under the roofs of houses.

Spreading disease

Houseflies are **pests** because they carry **disease**. A housefly has a hairy body, and when it walks over dirty surfaces or animal droppings, its hairs pick up dirt and germs. A fly constantly combs its hairs, dropping the dirt and germs on to food. This can spread diseases.

This fly is combing itself with its back legs. As it combs, germs and dirt will fall off its body.

Many flies, like these yellow dung flies, are attracted to dung. The flies may then land on food in kitchens.

People try to keep their kitchens free of flies by hanging up sticky fly papers which trap the flies. Or they use fly sprays which contain chemicals. The chemicals kill the flies within a few seconds.

This fly is sucking up sugar that it has found on a kitchen counter.

17

Maggots

Flies help to break down dead bodies. The smell of a rotting animal body attracts many flies. The flies crawl over the body and lay their eggs. Each egg hatches into a soft-bodied **larva** called a **maggot**.

A fly's eggs hatch within a few days. Soon there is a wriggling mass of maggots.

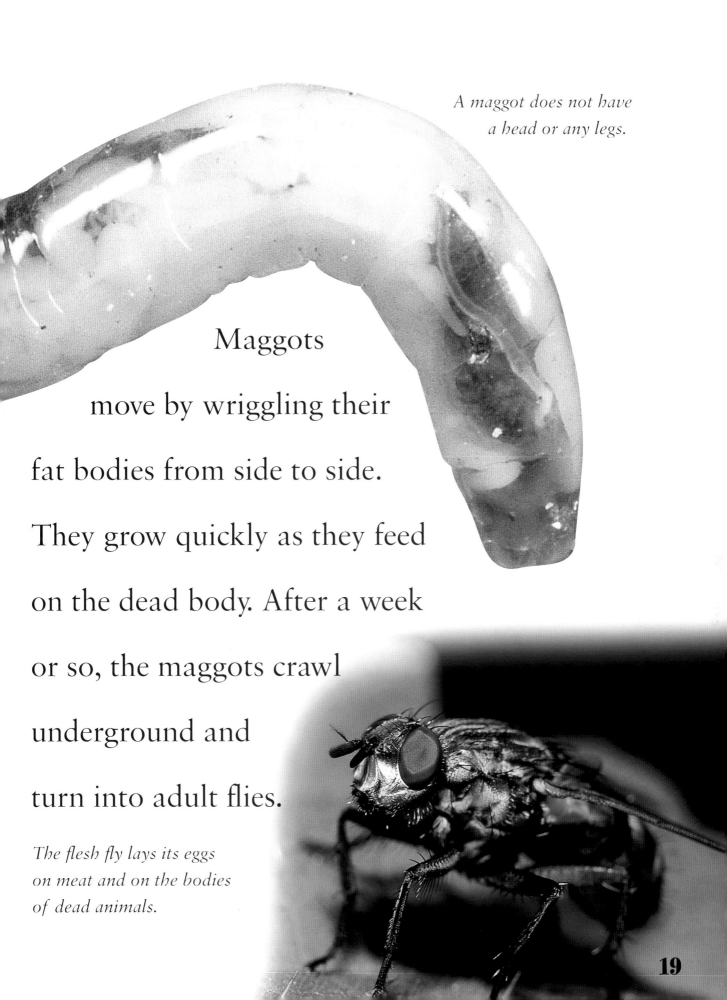

A maggot does not have a head or any legs.

Maggots move by wriggling their fat bodies from side to side. They grow quickly as they feed on the dead body. After a week or so, the maggots crawl underground and turn into adult flies.

The flesh fly lays its eggs on meat and on the bodies of dead animals.

Biting flies

Many flies are **carnivores**. This means that they feed on other animals. Robber flies and dung flies chase after smaller flies. They kill their **prey** and then suck out their body juices.

A yellow-bodied robber fly waits for a smaller insect to fly past. It grabs its prey with its strong legs and kills it.

The mouthparts of this cleg fly work like a syringe to suck up the blood.

Mosquitoes, midges and horseflies bite people. They all need a meal of blood before they can lay their eggs.

In some countries, female mosquitoes carry a disease called malaria. If an **infected** mosquito bites a person, it can pass on the disease.

The horsefly is found around horses and cattle. It can give you a painful bite.

Life cycles

Mosquitoes, midges and gnats live near water. These flies lay their eggs in small pools or ditches where the water is still and not moving. The eggs float on the surface of the water.

A female gnat lays her eggs in water. The eggs form small rafts which float on the surface of the water.

The eggs hatch into larvae which live in the water. These larvae can be seen hanging at the surface of the water where they can breathe air. Each larva turns into a **pupa**. A few days later, the pupal case splits open and a new adult comes out.

After it hatches, the larva (above) comes to the surface to breathe.

A new adult fly dries it wings before it flies away.

Looking like others

Many bees and wasps have brightly-coloured bodies. These are called warning colours because they warn other animals that they can sting. Some flies are the same colours as bees or wasps. This means that birds and other small animals that eat flies will leave them alone.

This yellow and black hover fly looks like a wasp. Birds avoid it because they think it can sting.

These drone flies look like honeybees. They visit flowers at the same time as bees.

There are hover flies that look just like wasps and honeybees. The bee fly even has a hairy body like a bee. You will see these flies when there are plenty of bees and wasps around.

The bee fly not only looks like a bee, it flies from flower to flower and sucks up nectar like a bee too.

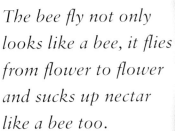

25

Flies with long legs

The fly with the longest legs is the crane fly, or daddy-longlegs. It has a long body too. Crane flies appear in the early evening. They are attracted by lights in homes. This large fly may look scary, but it is harmless.

A crane fly's long legs break off easily so that it can escape if it is caught by a predator.

The larva of a crane fly is called a leatherjacket. It lives in the soil and eats the roots of plants.

Some other long-legged flies have heavy bodies, so they fly slowly. As they fly through the air, their long legs hang down. Midges are small flies, and they have long legs too.

On a summer's evening, large groups of midges and gnats appear near water. Gnats make a buzzing sound, but midges are silent.

27

Watching minibeasts

Summer is the best time of year to look out for flies and mosquitoes.

Angling shops sometimes sell maggots as fish bait. You can watch how the maggots wriggle around. Try this simple test to see if maggots like the light. Place a large piece of paper on a table. Position a desk light so that it lights up half of the paper. Put a few maggots in the middle of the paper and see if they move towards or away from the light. Remember not to keep the maggots for too long or they will turn into pupae and become flies.

Watch how the maggots behave when they are placed under a light.

Gnats lay their eggs in water.

If you leave a small container of water outside during summer, it will attract gnats. The gnats will lay their eggs on the surface of the water. Within a few days you may be able to spot small animals swimming around in the water. These are probably gnat larvae.

Flies can carry germs that cause disease, so people do not like to have them in their homes. Some houses have screens or nets across the doors and windows to keep flies out. Fly sprays and sticky fly papers help to keep a home free of flies.

People cover food with cloths or nets to stop flies touching the food.

29

Minibeast sizes

Flies and mosquitoes are many different sizes. The pictures in this book do not show them at their actual size. Below you can see how big some of them are in real life.

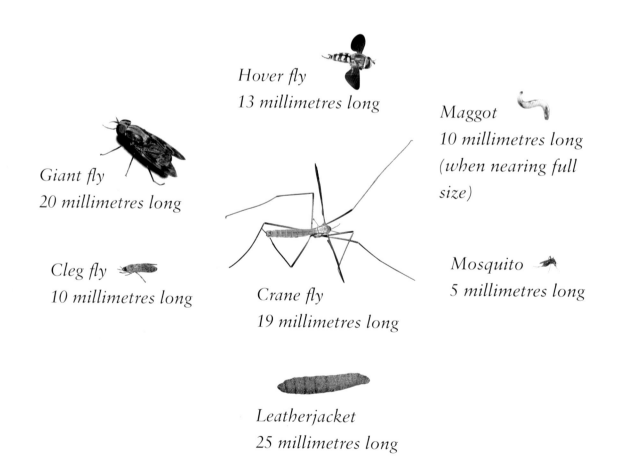

Hover fly
13 millimetres long

Maggot
10 millimetres long
(when nearing full
size)

Giant fly
20 millimetres long

Cleg fly
10 millimetres long

Crane fly
19 millimetres long

Mosquito
5 millimetres long

Leatherjacket
25 millimetres long

Glossary

abdomen The rear or back part of an insect's body.

antennae The feelers on an insect's head.

carnivores Animals that eat other animals.

disease An illness, usually caused by germs.

halteres Parts of a fly that look like knobbed stalks. They help a fly to balance while it is flying.

infected Carrying germs.

insects Animals with six legs and three parts to their bodies.

larva (plural: larvae) The young form of an insect. It looks different to an adult.

maggot The larva of a fly.

pests Animals that damage crops, homes, or carry germs and disease.

prey Animals that are killed by other animals for food.

proboscis (say: *pro-boss-iss*) A long mouthpart for piercing or sucking food.

pupa A hard case made from the skin of a larva. The larva turns into an adult inside the pupa.

Index

Editors: Claire Edwards, Sue Barraclough
Designer: John Jamieson
Picture researcher: Sally Morgan
Educational consultant: Emma Harvey

First published in the UK in 2001 by

Chrysalis Children's Books
An imprint of Chrysalis Books Group plc
The Chrysalis Building, Bramley Road
London W10 6SP
Paperback edition first published in 2003

ISBN 1 84138 351 1 (hb)
ISBN 1 84138 808 4 (pb)

Printed in Hong Kong
10 9 8 7 6 5 4 3 2 1 (hb)
10 9 8 7 6 5 4 3 2 1 (pb)

British Library Cataloguing in Publication Data
for this book is available from the British Library.

Picture acknowledgements:
Greenwood/Ecoscene: 24. Chinch Gryniewicz/Ecoscene:
25b. Wayne Lawler/Ecoscene: 26-27. Papilio: front &
back cover tcr & c, 5t, 5b, 18, 19b, 21t, 21b, 23t, 23b,
30clb. K.G. Preston-Mafham/Premaphotos: front cover
tr & cl & cr, 2, 3t, 4, 6, 7b, 9t, 10, 11tl, 12, 13t, 13b,
15t, 15r, 16, 17t, 20b, 22, 25t, 26b, 30cl, 30cb, 30crb.
M. Preston-Mafham/Premaphotos: 1, 14. R.A. Preston-
Mafham/Premaphotos: front & back cover tl & tcl, 7t,
11tr, 17b, 30c. Kjell Sanders/Ecoscene: 3b, 8, 9b.
Robin Williams/Ecoscene: 19t, 27t, 30cr, 30b.

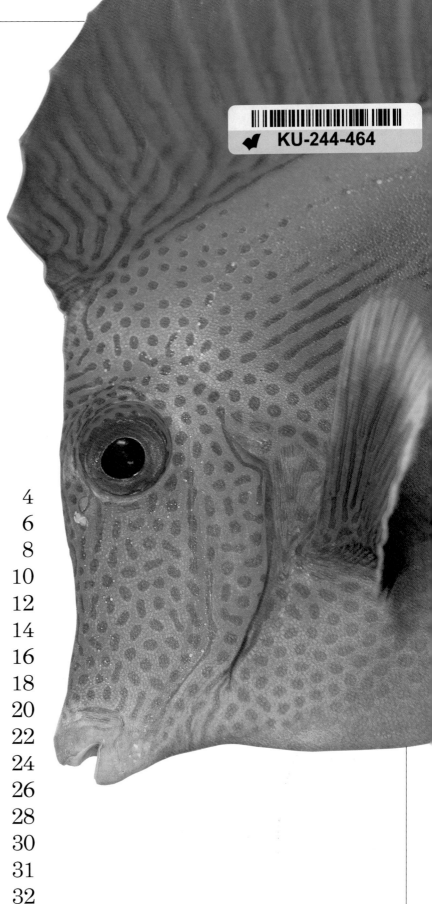

Contents

Meet the fish

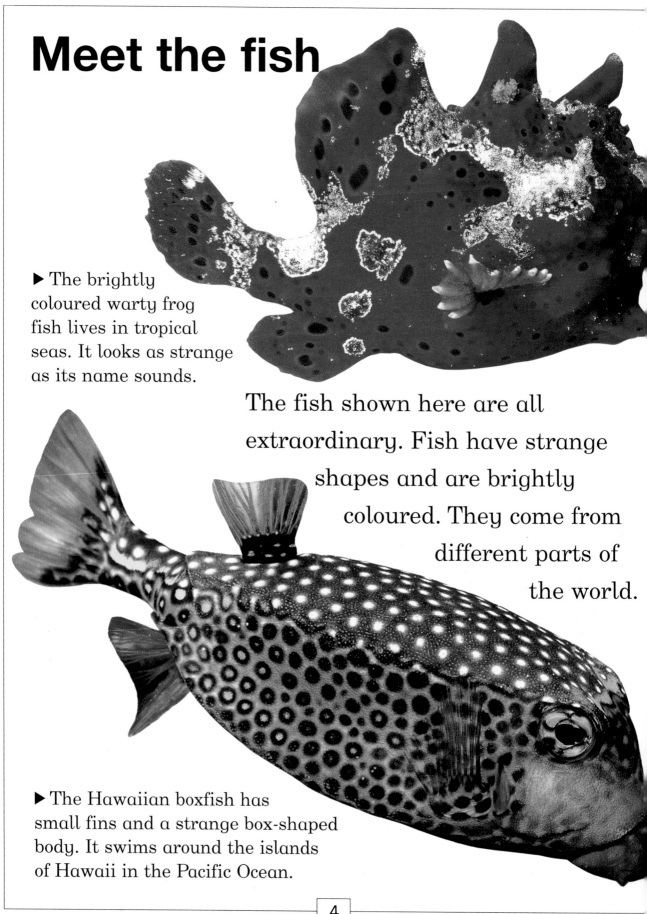

▶ The brightly coloured warty frog fish lives in tropical seas. It looks as strange as its name sounds.

The fish shown here are all extraordinary. Fish have strange shapes and are brightly coloured. They come from different parts of the world.

▶ The Hawaiian boxfish has small fins and a strange box-shaped body. It swims around the islands of Hawaii in the Pacific Ocean.

▼ The Red Indian fish has a long fin along its back like a Native American's head-dress. It lives off the coast of Australia.

All fish live in water – some in the sea, others in rivers and lakes. Many other animals, such as starfish, jellyfish and whales, live in water, too. Fish may not look alike, but they have some things in common that make them different from all other animals.

▶ The jigsaw trigger fish is also Australian. It lives on the world's largest **coral reef**, the Great Barrier Reef, off the east coast of Australia.

Types of fish

There are three main types of fish – jawless fish, sharks and rays, and bony fish. Jawless fish include lampreys and hagfish. They all have smooth, slimy skins.

Sharks and rays have rough, tough skins. Like jawless fish they have rubbery skeletons made of **cartilage**, not bone. Only the bony fish have hard bones and smooth **scales**.

◀ This trout has a river lamprey stuck to it. The lamprey's round mouth fixes on like a sucker while it feeds on the trout.

▲ A lamprey has no jaws. It uses its many tiny teeth to scrape away the flesh of its prey.

The Moorish idol and the trout are both bony fish. There are four times as many kinds of bony fish as other fish.

◀ A spotted wobbegong shark lurks on the seabed. It does not look like a shark, but it has rough skin and a rubbery skeleton.

Breathing in water

All living things need to breathe in **oxygen**. Land animals, including humans, take oxygen from the air they breathe into their **lungs**.

▼ Fish take in oxygen through their gills. This reef shark has four gill slits behind its head, just in front of its fins.

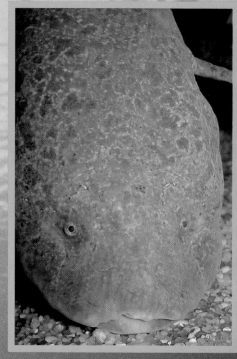

◀ Lungfish live in lakes and marshes. They have gills and lungs. They use their gills under water. If the water dries up, they breathe in air through their lungs.

Most fish have gills instead of lungs. A fish swallows water through its mouth and pushes it out through its gills. As water passes over the gills, oxygen from the water passes into the fish's blood. At the same time **carbon dioxide** passes from the blood into the water.

▶ Bony fish have gill flaps that cover and protect their gills. The flap opens to let the water through.

Did you know?

Most fish open and shut their mouths to breathe, but some sharks just swim along with their mouths open.

▼ The side of this bony fish has been cut away to show the gills which are behind its eyes. The gills take in oxygen from the water as it flows over them.

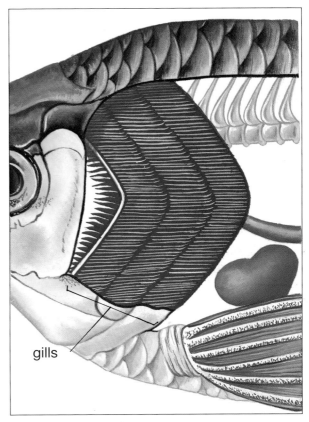

gills

Moving through water

A fish has very powerful **muscles** down the sides of its body. It uses them to move its tail from side to side to swim through the water. The blue-finned tunny can swim at over 60 kilometres an hour – seven times as fast as an Olympic swimmer.

▶ A bony fish has an extra **organ** – a swimbladder – inside its body. The swimbladder is filled with gas and allows the fish to move up and down in the water.

▼ Close up of the swimbladder in the fish below.

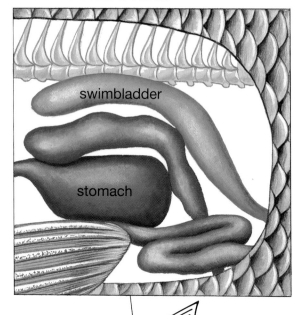

swimbladder

stomach

▶ Salmon can leap up and over a waterfall. They swim to the surface and flick their tails to glide through the air.

A bony fish has a swimbladder filled with gas to stop it sinking in the water. When the fish wants to rise, it lets more gas into the swimbladder. When it wants to sink deeper, it lets some gas out.

▼ Fish swim by moving their tails. The pickerell's long, thin shape helps it to swim fast through the water.

Fins

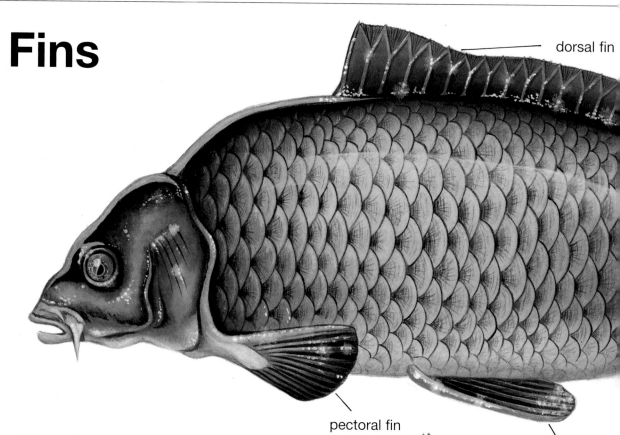

dorsal fin

pectoral fin

pelvic fin

All fish have fins. Their tail fin and muscles push them through the water. Their other fins help them to stop and steer, and keep them steady. If a fish wants to move slowly, it flaps its fins.

Some fish use their fins for other things too. Mudskippers sometimes leave the water to look for food. Then they use their fins like feet.

◀ A seahorse is a fish, although it doesn't look much like one. It has very small fins that move it slowly through the water.

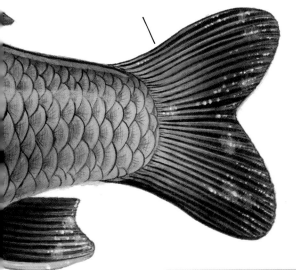

caudal fin

anal fin

Did you know?

The fin on the back of the deep-sea anglerfish stretches over its head. It has a light on the end to attract **prey.**

▼ A flying gurnard has huge side fins. It uses them like wings when it leaps from the water and glides through the air.

◀ The lionfish has brightly coloured fins and poisonous spines. The large, waving fins warn other animals to stay away from the spines and not to try to eat it.

Seeing underwater

Most fish have big, bulging eyes on both sides of their heads. They can see all around and above and below at the same time. They look out for food and danger.

▶ A fish's eye is like a human eye, except that the **lens** is round and bulges through the **pupil**.

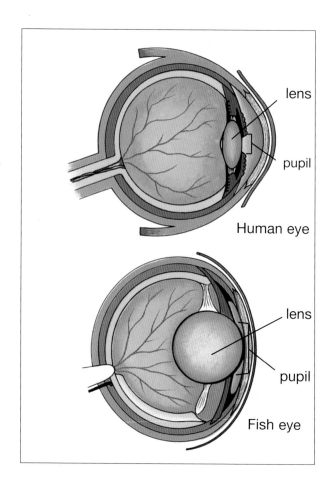

lens

pupil

Human eye

lens

pupil

Fish eye

▲ Mudskippers spend a lot of time sitting in shallow water or mud.

▶ A deep-sea hatchet fish has very large eyes. They help the fish to see in the dark depths of the ocean.

Mudskippers have eyes on top of their heads to see insects above the water. Some fish can see in very little light. The deeper they go in the sea the darker it becomes.

◀ The black sea dragon lives in total darkness in very deep sea. It makes its own light to attract fish into its huge, gaping jaws.

Making sounds

Sound travels better through water than it does through air. A fish does not need ears outside its head to hear underwater. Instead its ears are sealed inside its head, behind the eyes. Sounds travel through its body to the ears.

Fish can hear sounds many kilometres away. Many fish can make sounds, too. Some grind their teeth together, some rub their fins against their bodies. Many fish use their swimbladders to make a noise and to help them hear better.

▶ You cannot see a fish's ears because they are covered up inside its body.

▲ This catfish makes a noise with its swimbladder. Special muscles **vibrate** the swimbladder like a drum. The swimbladder also helps the fish to hear. Sound makes it vibrate like an **eardrum**.

▶ Loaches squeak like mice. They do this by blowing gas out of their swimbladders. You can make a similar noise by taking a blown-up balloon and pulling the mouth flat as you let the air out.

Did you know?

Many fish are very noisy eaters. Toad fish sound almost as loud as an underground train. Their noises warn other toad fish to keep away from their food.

Smelling and tasting

Smelling and tasting are more important to fish than to humans. A sharks has a very good sense of smell – more than half its brain is devoted to it. We taste with our tongues, but a fish has taste buds all around its head and in its mouth. It has **taste buds** on its body and front fins, too. Catfish and cod can taste with their whisker-like feelers.

◀ Sharks can smell blood up to two kilometres away. They quickly gather from all around when an animal is killed or wounded.

Did you know?

A salmon is born in a river then swims out to sea, but it always returns to the same river to breed. It finds the river by smell and by taste.

A special sense

Some fish have an extra, special sense. Along each side of their bodies they have a line, called the lateral line, which can detect the smallest movement in the water around them.

▶ This perch has a lateral line on its side.

The lateral line stops a fish swimming into rocks, or into the sides of a fish tank. It allows large groups, or shoals, of fish to swim close together. If the shoal is attacked, the fish swim off in all directions without colliding.

▼A shoal of sardines swims around a rock. Each fish changes direction and avoids the rock without bumping into any of the others.

Feeding

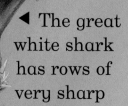

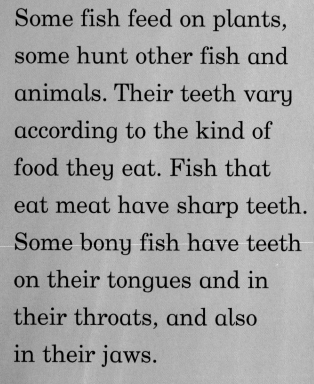

◀ The great white shark has rows of very sharp teeth. When one tooth falls out it is immediately replaced by another.

Some fish feed on plants, some hunt other fish and animals. Their teeth vary according to the kind of food they eat. Fish that eat meat have sharp teeth. Some bony fish have teeth on their tongues and in their throats, and also in their jaws.

◀ An angel fish feeds on hard coral. Its teeth have joined together into a sharp beak which cuts through the coral.

22

gill-rakers

Very often (but not always) big fish eat little fish, which feed on still smaller fish, which feed on plants. One of the very biggest fish, the basking shark, also feeds on plants. Its mouth has a special strainer called a gill-raker.

▲ A basking shark feeds on tiny plants and animals called plankton. It has a gill-raker in its mouth which strains the plankton from the water.

▶ Goldfish live in rivers and feed off the plants that grow on the bottom and sides of the river.

Defences

As well as finding food, fish try to avoid being eaten. Some fish, such as toad fish, stonefish, and weever fish, defend themselves with poisonous spines. The anemone fish defends itself by hiding in the poisonous tentacles of the sea anemone. Its slimy skin stops it being stung itself.

▲ The stonefish (at the top) has two defences. It has poisonous spines and it looks like a stone.

▲ The sargassum fish (in the centre) looks just like the seaweed it swims among.

◀ The anemone fish can swim safely among the stinging tentacles of the sea anemone.

Many fish defend themselves by looking like their surroundings. Fish that swim in the ocean are often dark on top, to look like the sea, and silvery below, to look like the sky.

► The leaf-fish looks like a dead leaf floating in the water. It is hard to make out its head and fins.

▲ When attacked, the spiny pufferfish swallows huge amounts of water to puff out its sharp, strong spines.

Creating young

It takes a male and a female fish to produce new fish. The female lays her eggs and the male **fertilizes** them with his sperm.

Most fish produce thousands of eggs. They leave them to float in the ocean, and many are eaten by other fish.

A few fish, such as seahorses, look after the eggs until they hatch. Sharks' eggs are fertilized inside the female and are born already hatched.

The male seahorse looks after the fertilized eggs in a pouch on his belly.

▼ As the female brown trout lays her eggs, the male fertilizes the eggs with his sperm.

▲ A young rainbow trout hatching from its egg. The black spots are unhatched fish.

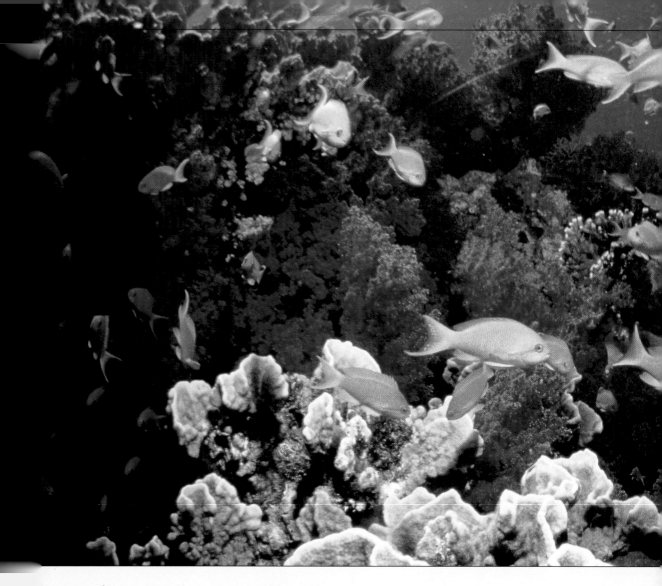

World of fish

Water covers nearly three-quarters of the Earth, and more than 20 000 different fish live in it. Some live in warm, tropical seas, especially around coral reefs.

But many fish prefer the cold water of the Arctic and Antarctic Oceans. Freshwater fish live in rivers, lakes and streams. Each fish is well suited to the place it lives.

Colourful fish
swimming
around coral
in the Red Sea.

Glossary

Carbon dioxide A gas that is carbon and oxygen combined together. It is one of the gases found in the air.

Cartilage A tough, rubbery substance that sharks and some other fish have instead of bones. Human ears are also made of cartilage.

Coral reef This looks like rock, but is the skeletons of countless tiny relatives of sea anemones.

Eardrum A thin skin stretched across the ear canal, which vibrates when sound waves reach it. The vibrations are passed through the ear to the brain.

Fertilizes When a male sperm joins with a female egg it fertilizes the egg. An egg must be fertilized before it can grow into a baby fish.

Fins Parts of a fish which stick out from its body and help it to move through the water.

Lens The part of the eye which focuses the light to give a clear picture.

Lungs Part of the body where oxygen from the air passes into the blood.

Muscle Meaty substance inside the body which makes the bones and other parts of the body move.

Organ A part of the body with a particular job to do. Animals have many kinds of organs, for example, a heart, stomach and eyes. Most bony fish also have a swimbladder.

Oxygen A gas which plants and animals need to stay alive. Oxygen is in air and in water.

Prey An animal hunted for food.

Pupil The black hole in the middle of the eye. Light goes through this into the eye.

Scales Small, flat plates that cover the skin and protect it.

Taste buds Special cells which react to chemicals and produce the sense of taste.

Vibrate Move backwards and forwards or up and down very fast.

Key facts

Largest fish The whale shark is the largest fish. It grows up to 18 metres long, nearly as long as two buses parked end to end.

Largest freshwater fish The beluga is a kind of sturgeon that lives in the Russian Federation. It grows up to 8.4 metres long, nearly as long as three cars. Its eggs provide the best caviar, a very expensive food.

Smallest fish The pygmy goby grows only about 12 mm long, about as long as your thumbnail.

Fastest fish A sailfish can reach 109 kilometres an hour. If a sailfish competed in the Olympics, it could swim nearly 12 lengths before the fastest human could swim one.

Most dangerous shark Great white sharks have attacked more humans than any other kind of shark, but tiger sharks are known for the fierceness with which they eat almost anything.

Most poisonous fish Stonefish probably have the most poisonous spines of any fish.

Most eggs laid The ocean sunfish lays up to 300 million eggs at a time. It lives in the sea and most of the eggs and young fish are eaten by other animals. Only one or two young sunfish survive to become adults.

Fewest eggs The stickleback lays only a few eggs. To protect the eggs, the male builds a nest of roots and leaves cemented into a tube with a special liquid. He then chases the female into the tube, where she lays the eggs. The male guards the fertilized eggs until they hatch.

Coldest fish Icefish live in the Antarctic Ocean where the water is often less than 0 degrees Centigrade. They have their own anti-freeze to stop their blood from freezing solid.

Earliest fish The first known fish were jawless fish called ostracoderms. They lived about 510 million years ago, and were covered in armour. The first bony fish were spiny sharks called acanthodians. They lived about 410 million years ago.

Index